Helen Hunt Jackson

The procession of flowers in Colorado

Helen Hunt Jackson

The procession of flowers in Colorado

ISBN/EAN: 9783337109301

Printed in Europe, USA, Canada, Australia, Japan

Cover: Foto ©berggeist007 / pixelio.de

More available books at **www.hansebooks.com**

THE PROCESSION OF FLOWERS IN COLORADO.

SUPPOSE the little black boys who hang on lamp-posts along the route of a grand city procession are not the best reporters of the parade. They do not know the names of the officials, and they would be likely to have very vague ideas as to the number of minutes it took the procession to pass any given point; but nobody in all the crowd will have a more vivid impression of the trappings of the show, of the colors and the shapes, and of the tunes the bands played. I am fitted for a chronicler of the procession of flowers in Colorado only as little black boys are for chroniclers of Fourth of July processions. Of the names of the dignitaries, and the times at which they reached particular places, I am sadly ignorant; but there

7

is hardly a color or a shape I do not know by sight and by heart, and as for the music of delight which the bands play, its memory is so vivid with me that I think its rhythm would never cease to cheer me if I were banished for ever to Arctic snows.

The first Colorado flower I saw was the great blue windflower, or anemone. It was brought to me one morning, late in April, when snow was lying on the ground, and our strange spring-winter seemed to be coming on fiercely. The flower was only half open, and only half way out of a gray, furry sheath some two inches long ; it looked like a Maltese kitten's head, with sharp-pointed blue ears, — the daintiest, most wrapped-up little blossom. " A crocus, out in chinchilla fur ! " I exclaimed.

" Not a crocus at all ; an anemone," said they who knew.

It is very hard, at first, to believe that these anemones do not belong to the crocus family. They push up through the earth in clusters of conical, gray, hairy buds, and open cautiously, an inch or two from the ground, precisely as the crocuses do ; but day by day, inches at a time, the

8

stem pushes up, until you suddenly find some day in a spot where you left low clumps of what you will persist, for a time, in calling blue crocuses, great bunches of waving blue flowers, on slender stems from six to twelve inches high, the blossoms grown larger and opened wider, till they look like small tulip cups, like the Italian anemones. A week or two later you find at the base of these clumps a beautiful fringing mat of leaves, resembling the buttercup leaf, but much more deeply and numerously slashed on the edges. These, too, grow at last away from the ground, and wave in the air ; and by the time they are well up, many of the flowers have gone to seed, and on the top

9

of each stem flutters a great ball of fine, feathery seed plumes, of a green or claret color, almost as beautiful as the blossom itself. These anemones grow in great profusion on the foot-hills of the mountains to the west of Colorado Springs. They grow even along the road-sides, at Manitou. They have, apparently, caprices of fondness for certain localities, for you shall find one ridge blue with them, and another, near by, without a single flower.

About the same time as the anemone, or a little before, comes the low white daisy, harbinger of spring in Colorado, as is the epigæa in New England. This little blossom opens at first, like the anemone, close to the ground, and in thick-set mats, the stems so short, you can get the flower only by uprooting the whole mat. It has a central root like a turnip, from which all the mats radiate, sometimes a dozen from one root. Take five or six of these home, and fill a low dish with them, and the little brown blades of leaves will freshen and grow up like grass, and the daisies will peer up higher and higher, until the dish looks like a bit of a waving field of daisies.

Flowers in Colorado.

Next after these comes the mountain hyacinth, popularly so called for no other reason than that its odor is like the odor of the hyacinth. It is in reality a lily. It is the most ethereal and delicate of all our wild flowers, and yet it springs up, like the commonest of weeds, in the commonest of places; even in the dusty edges of the streets, so close to the ruts that wheels crush it, it lifts its snowy chalice. On neglected opens, in pathways trodden every day, you may see these lilies by dozens, trampled down; and yet at first sight you would take them for rare and fragile exotics. The blossom is star-shaped, almost precisely like the white jessamine, and of such fine and transparent texture that it is almost impossible to press it; one, two, sometimes half a dozen flowers, rising only two or three inches high from the centre of a little bunch of slender green leaves, in shape like the blades of the old-fashioned garden-pink, but of a bright green color. It is one of the purest looking blossoms. To see it as we do, growing lavishly in highways, trodden under foot of man and beast, is a perpetual marvel which is never quite free from pain.

After these three forerunners comes a great outburst
of flowering : yellow daisies of several varieties, yellow
mustard, a fine feathery, white flower, and vetches of all
sizes, shapes, colors, more than you can count. And here
I am not speaking of what happens in nooks and corners
of the foot-hills, in fields, or by-ways, or places hard to
come at. I am speaking of what happens in the streets of
Colorado Springs, along all the edges of the sidewalks, in
little spaces left at crossings, in unoccupied lots, — in
short, everywhere in the town where man and his houses
have left room. It is not the usual commonplace of exag-
geration ; it is only the simplest and most graphic form of
exact statement you can find, to say that by the middle
of June the ground is a mosaic of color. The vetches are
bewildering. There are sixteen varieties of vetch which
grow in one small piece of table-land between the Colorado
Springs Hotel and the railroad station. They are white,
with purple markings, all shades of purple, and all shades
of red ; some of them grow in spikes, standing erect ;
some in scrambling and running vines, with clusters of
flowers ; some with single blossoms, like the sweet-pea,

12

and as varied in color. They all lie comparatively low, partly from the want of bushes and shrubs to climb on, partly because they are too wise to go very far away from their limited water supply in so dry a country; they must keep close to the ground, or choke. That this is a bit of specific precaution on their part, and not a peculiarity of their varieties, is proved by the fact that all along the creeks, in the cotton-wood and willow copses, we find the same vetches growing up boldly many feet into the air, just as they do in Italy, leaping from shrub to shrub, and catching hold on anything which comes to hand.

By the third week in June, we have added to these brilliant parterres of red, purple, white, and yellow in our streets the superb spikes of the blue pentstemon. This is a flower of which I despair to give any idea to one a stranger to it. The blossoms are shaped like the common foxglove blossom; they grow

on the stems in single, double, or triple rows, as may be. I have seen stems so tight packed with blossoms that they could not stand erect, but bent over, like a bough too heavily loaded with fruit. Before the blue pentstemon opens, it is a delicate pink bud ; when it first opens, it is a clear, bright blue, as blue as the sky ; day by day its tints change, sometimes to a purplish-blue ; sometimes back again towards its childhood's pink, so that out of a hundred spikes of blue pentstemon you shall see no two of precisely the same tint ; when they are their deepest, most purple blue, they look like burnished steel ; when they are at their palest pink, they are as delicate as a pink apple-blossom. O New Englander ! groping reverently among scattered sunny knolls and in moist wood depths for scanty handfuls of pale blossoms, what would you do at such a banquet as this, spread before you whenever you stepped outside your door, lying between you and the post-office every day? For, let me repeat, these flowers of which I have spoken thus far are only the flowers which grow wild in our streets, and there are yet many that I have not mentioned : there is the dark blue spider-wort, which is everywhere ;

and there are several yellow flowers, and one of pale pink, and several of white, I recollect, whose names I do not know; neither do I know how to describe their shapes. I am as helpless as the little black boy on the Fourth of July, — I can describe only the colors.

Leaving the streets of the town, and going southwest towards the foot-hills of the Cheyenne Mountain, we come to a new and a daintier show. As soon as we strike the line of the little creek which we must follow up among the hills, we find copses of wild plum and wild roses in full bloom. The wild rose grows here in great thickets, as the black alder grows in New England swamps. The trees are above your head, and each bough is so full of roses it would seem an impossibility for it to hold one rose more. We bear wild roses home, by whole trees, and keep them in our rooms in great masses which will well-nigh fill a window. I have more than once tried to count the roses on such a sheaf in my window, and have given it up.

Along the banks of the brook are white daisies and pink; vetches, and lupines, white, yellow, and purple. The yellow ones grow in superb spikes, one or two feet from the ground ; and the white ones in great branching plants, six or seven from a single root. On the first slopes of the foot-hills begins the gilia. This is a flower hard to describe. Take a single flower of a verbena cluster; fancy the tubular part an inch or two long, and the flowers set at irregular intervals up and down the length of a slender stem; this is the best my ignorance can do to convey the idea of the shape of the gilia. And of the color, all I can say is that the gilia is what the botanists call a sporting flower; and I believe there is no shade of red, from the brightest scarlet up through pale pinks, to white, which you may not see in one half an acre where gilias grow. It is a dancing sort of flower, flutters on the stem, and the stem sways in the lightest wind, so that it always seems either coming towards you or running away.

There is a part of Cheyenne Mountain which I and one other have come to call " our garden." The possessive pronoun has no legal title behind it; it is an audacious assumption not backed by any squatter sovereignty, nor even by any contribution towards the cultivation of the soil; but ever since we found out the place, it has

16

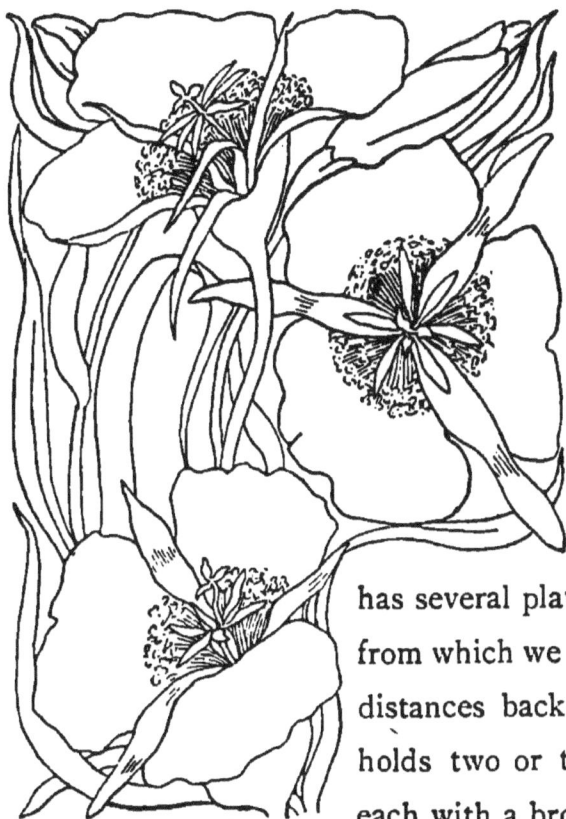

been mysteriously worked " on shares " for our benefit ; and as long as we live we shall call it our garden. It lies five or six hundred feet above the town, four miles away, and has several plateaus of pine groves from which we look off into eastern distances back of the sunrise; it holds two or three grand ravines, each with a brook at bottom ; it is walled to the west by the jagged and precipitous side of the mountain itself. The best part of our " procession of flowers " is always here.

Here on the plateaus, under the shade of the pines, are the anemone in stintless numbers, daisies, and kinnikinnick.

17

In June the kinnikinnick vines are full of little pinkish-white bells, shaped like the wintergreen bell, and as fragrant as the linnæa blossom. Here are three low-growing varieties of the wild rose, none more than two or three inches from the ground: one pure white, one white with irregular red markings, and one deep pink. The petals are about one-third larger than those of the common wild rose.

Here are blue violets, and in moist spots the white violet with a purple and yellow centre. Here is the common red field lily of New England, looking inexplicably away from home among pentstemons and gilias, as a country belle might in court circles. Here is the purple clematis ; a half-parasitic plant this seems to be, for you find it wound up and up to the very top of an oak or cherry bush, great lengths of its stem looking as dead as old drift-wood, but whorls of lovely, fringing green leaves and purple, cup-shaped blossoms bursting out at intervals, sometimes a foot apart. How sap reaches them, through the cracked and split stems, it is hard to see ; but it does, for you can carry one home, trellis and all, set it in water, and the clematis will live as long as the oak bush will.

18

Flowers in Colorado.

Here is the purple pentstemon, never but a single row
of blossoms on its stem, and the scarlet pentstemon, most
gorgeous of its family. This, too, has but a single row of
flowers on its stem ; they are small, of the brightest scarlet,
and the shape is somewhat different from the other pentstemons ; longer, slenderer, and more complicated, they look
like fairy gondolas hung by their prows. I have seen the
stems as high as my shoulder, and the scarlet gondolas
swinging all the way down to within a foot of the ground.

Here are great masses of a delicate flowering shrub, a
rubus, I think I have heard it called. Its flower is like
a tiny single-petalled rose, of a snow-white color. On first
looking at the bush, you would think it a wild white rose,
till you observed the leaf, which is more like a currant
leaf. Here also are bushes of the Missouri currant, with
its golden-yellow blossoms, exhaustless in perfume ; and a
low shrub maple, which has a tiny, apple-green flower, set in
a scarlet sheath close at the base of each leaf, so small
that half the world never discovers that the bush is in
flower at all. Here are blue harebells, and Solomon's seal
both low and high; and here is the yellow cinquefoil. In
the moist spots, with the white violets, grows the shooting

star, finer and daintier than the Italian cyclamen; its sharp-pointed petals of bright pink fold back like rosy ears; in its centre is a dark-brown circle round a sharp needle-point of yellow. There are many more, but of all the rest I will speak only of one, — the great yellow columbine. This grows in the ravines. The flower is like our garden columbine, but larger, and of an exquisite yellow, sometimes with white in the centre. It grows here in such luxuriant tufts and clumps that you will often find thirty and forty flower-stems springing up from one root. Of this plant I recollect the botanical name, which was told me only once, but I could no more forget it than, if I had once sat familiarly by a queen in her palace, I could forget the name of her kingdom. It is the golden columbine of New Mexico, the aquilegia chrysantha.

When we drive down from "our garden" there is seldom room for another flower in our carriage. The top thrown back is filled, the space in front of the driver is filled, and our laps and baskets are filled with the more delicate blossoms. We look as if we were on our way to the ceremonies of Decoration Day. So we are. All June days are decoration days in Colorado Springs, but it is the

sacred joy of life that we decorate,—
not the sacred sadness of death. Going
northwest from the town towards the
mesa or table-land which lies in that
direction between us and the foot-
hills, we find still other blossoms, no
less beautiful than those of which I
have spoken: the wild morning-glory
wreathes the willow bushes along the
Fountain Creek which we must cross,
and in the sandy spots between the
bushes grow the wild heliotrope in
masses, and the wild onion, whose
delicate clustered umbels, save for their odor,
would be priceless in bouquets. Yellow lu-
pine, red gilias, wild roses, and white spiræas
are here also; and waving by the roadsides, careless and
common as burdocks in New England, grows the superb
mentzelia. This is a regal plant; the leaves are of a bluish-
green, long, jagged, shining, like the leaves of the great
thistles which so adorn the Roman Campagna; the plant
grows some two feet or two and a half feet high, and

branches freely. Each branch bears one or more blossoms,
— a white, many-pointed starry disk, in its centre a wide
falling tuft of fine silky stamens. Here also we find a
large white poppy whose leaves much resemble the leaves
of the mentzelia ; and in the open stretches beyond the
creek, the ground is white and pink every afternoon with
the blossoms of four-o'clocks. There must be several
varieties of these, for some are large and some are small,
and they have a wide range of color, white, pinkish-white,
and clear pink. Higher up, on the top of the *mesa*, we
come to great levels which are dotted with brilliant points
of fiery scarlet everywhere ; the first time one sees a
scarlet "painter's brush" (castilleia) a few rods ahead of
him in the grass is a moment he never forgets ; it looks
like a huge dropped jewel or a feather fallen from the
plumage of some gorgeous bird. There are two colors
of the castilleia here, — one, of an orange shade of scarlet,
and the other of the brightest cherry red. But beautiful
as is the castilleia, it is not the *mesa's* crowning glory :
vivid as is its color, the pale creamy tints of the yucca
blossoms eclipse it in splendor. This also is a thing a
lover of flowers will never forget, — the first time he saw

yuccas by the hundred in full flower out-of-doors. It grows in such abundance on this *mesa* that in winter the solid green of its leaves gives a tone of color to whole acres. Spanish bayonet is its common name here, and not an inappropriate one, for the long, blade-like leaves are stiff and pointed as rapiers. They grow in bristling bunches directly from the root ; the outer ones spread wide, and sometimes lie on the ground ; from the centre of this "chevaux de frise ' rise the flower-spikes, usually only one, sometimes two or three, from one to two and a half feet high, set thick with creamy white cups which look more like a magnolia flower than like any thing else. I counted once seventy-two on a spike about two feet long. Profusely as the yucca grows on this *mesa*, we do not get so many of them as we would like, for the cows are fond of them and eat the blossoms as fast as they come out. What a picture it is, to be sure, — a vagrant cow rambling along mile after mile, munching the tops of spikes of yucca blossoms ! There ought to be something transcendent in the quality of her milk after such a day as that.

Beside the castilleia and the yucca, there grow on this *mesa* many of the vetches, especially a large white variety, which I have a misgiving that I ought to call astragalus, and not vetch.

The *mesa* slopes away to the east and to the west ; it is really a sort of causeway, or flattened ridge; on its sides are innumerable small nooks and hollows which, catching and holding a little more moisture than the surface above, are full of oak-bushes, little green oases on the bare slopes; in these grow several flowering shrubs, spiræas, and others whose names I know not.

Crossing the *mesa* and entering the foot-hills again, we come to little brook-fed glens and parks where grow all the flowers I have mentioned ; yes, and more, for, I bethink me, I have not yet spoken of the white clematis, — virgin's bower, as it is called in New England. This runs riot along every brook-course in the region, — this and the wild hop, the white feathery clusters of the one and the swinging green tassels of the other twisting and intertwisting, and knitting everything into a tangle ; and the blue iris also, in great spaces in moist meadows, and the dainty nodding bells of the wild flax a little farther up on the hills, and the yellow lady's-slipper, and the coreopsis, and the mertensia, which has drooping spikes of small blue bells

24

that are pink on the outside when they are folded up. And
I believe that there are yet others which I do not recollect,
besides some which I remember too vaguely to describe,
having seen them perhaps only once from a car window,
as I saw a gorgeous plant on the Arkansas meadows,
one day. It was a great sheaf of waving feathery spikes
of yellow. It is true that a railroad train waited for me
while I had this plant taken up and brought on board. I
nursed it carefully with water and shade all the way from
Pueblo to Colorado Springs, but it was dead when I
reached home, and nobody could tell me its name. After-
wards a botanist told me that it must have been stanleya
pinnatifida, but I liked my name for it better,—golden
prince's feather.

If it were ever possible to weary of the flora in the
vicinity of Colorado Springs, and to long for some new
flowers, one need but go a few hours farther south to
Canyon City, and he will strike an almost tropical flora.
Here grow twelve different varieties of cactus either in the
town itself or on the slopes of the hills around it ; some of
these varieties are very rare ; all bear brilliant blossoms,
yellow, scarlet, and bright purple. Here grow all the

flowers which we have at Colorado Springs, with many others added.   A short extract from a paper written by an enthusiastic Canyon City botanist will give to botanists a better idea of the flora of Colorado than they could get from volumes of my rambling enthusiasm.

" There is no pleasanter botanical trip in the vicinity of Canyon City than a walk beyond the bath-rooms of the hot springs to the gate of the mountains, up the canyon of the Arkansas, and to the top of the Grand Canyon, a distance of about four miles.   The grandeur of the far mountain summits covered with eternal snow, the perpendicular cliffs over one thousand feet high, the great river boiling and dashing along its rocky channel, are sources of excitement nowhere else combined ; but to any one interested in flowers, their beauty, their abundance, and the rare species that meet you at every step make the trip wonderfully interesting.   Here among the rocks are the most northern known stations of the ferns, pellæa wrightiana and cheilanthes eatoni; and on the walls of the Grand Canyon, more than a thousand feet above the river, grows the very rare asplenium septentrionale, which the wild big-horn, or mountain sheep,

26

seem to appreciate so much that it is difficult to find a specimen not bitten by them. The syringa (philadelphus microphyllus) is growing wherever it can find a foot-hold, and here and there is a bunch of the rare western Emory's oak, that, like several other plants, seems to have wandered in from the half-explored region of the great Colorado River of Arizona. The lateral canyons are full of fallugia paradoxa, with its white flowers and plumed fruit, and where little streams of water come dashing over the rocks and losing themselves in mist, the golden columbine of New Mexico, aquilegia chrysantha, grows to perfection. The scarlet pentstemon, blue pentstemon, the brilliant gilia aggregata, spiræas, castilleias, and hosts of less showy but equally interesting plants occupy every available piece of soil. The beauty of the flora is as indescribable as the grandeur of the scenery.

" The abundance of the four-o'clock family is noticeable. All of the nyctaginaceæ of Colorado are found about Canyon City, and some of them as yet only in this part of the territory. Most of them are very interesting, and their beauty forms a very prominent feature of our flora in June and July. Abronia fragrans whitens whole acres of land ;

and the large, conspicuous flowers of mirabilis multiflora are seen all over the town ; opening their flowers late in the afternoon in company with the vespertine mentzelias, they are fresh and bright during the most pleasant part of the summer day. The Soda Spring Ledge, from which boils the cold mineral water, is a locality rich in rare plants. Here grow thamnosma texana, abutilon parvulum, allionia, incarnata, tricuspis acuminata, mirabilis oxybaphoides, etc.

" The common flowers of Colorado are very abundant around Canyon City and in its vicinity. The monarda grows upon the *mesas ;* exquisite pentstemons adorn the brooks ; rosa blanda and the more beautiful rosa arkansana are found on the banks of the Arkansas ; eriogonum and astragalus are numerous in species and numberless in specimens ; the grass fields of Wet Mountain Valley are full of clovers and cypripedium, iris and lilies ; the botanist, wandering through the canyons of the Sangre di Cristo range, tramples down whole fields of white and blue larkspur and delicate mertensia. The summits are covered with woolly-headed thistles, phlox, senecios, forget-me-nots, saxifraga, and the numberless beauties of the Alpine flora. And besides all this, perhaps no locality in the world affords

28

Flowers in Colorado.

better opportunities to the collector to fill his herbarium with beautiful and rare specimens easily and rapidly. The wealth of foliage found in moister climates does not obstruct the view and hide the more modest flowers, while the perpendicular range of nearly two thousand feet, through which he may pass on his botanical rambles, carries him from a climate as genial as that of Charleston to one as thoroughly boreal as that of the glaciers of Greenland." Not the least of the delights of living in such a flower-garden as Colorado in June and July is the delight of seeing the delight which little children take in the flowers. Whenever in winter I try to recall the face of our June, I think I recall the blossoms oftenest as they look in the hands of the school-children. Morning, noon, and evening, you see troops of children going to and fro, all carrying flowers ; the babies on doorsteps are playing with them ; and late in the

afternoon, as you drive through the streets, you see many a little sand-heap in which are stuck wilted bunches of flowers, that have meant a play-garden all day long to some baby who has gone to sleep now, only to wake up the next morning and pick more flowers to make another garden. And among the sweet sayings which I have heard from the mouths of children, one of the very sweetest was that of a little girl not six years old, who has never known any summer less lavish than Colorado's. As soon as the flowers come she is impatient of every hour she is obliged to spend in-doors. At earliest dawn she clamors to be taken up and dressed, exclaiming, " I must get up early, there is so much to do to-day ; there are so many flowers to be picked." Coming in one day with her hands full of flowers which had grown near the house, she gave them one by one to her mother, gravely calling them by their names as she laid them in her mother's hand. Of the last one, a tiny blue flower, she did not know the name. Looking at it earnestly for a moment or two, she said hesitatingly, as she placed it with the rest, "And this one — this — is a kiss from the good God. He sends them so."

30

HELEN JACKSON'S WRITINGS.

A KEY TO "RAMONA."

A CENTURY OF DISHONOR.

A Sketch of the United States Government's Dealings with some of the Indian Tribes.

A New Edition. 12mo. pp. 514. Cloth. $1.50.

Mrs. Jackson devoted a whole year of her life to writing and compiling materials for "A Century of Dishonor," and while thus engaged she mentally resolved to follow it with a story which should have for its *motif* the cause of the Indian. After completing her "Report on the Condition and Needs of the Mission Indians of California" (see Appendix, p. 458) she set herself down to this task, and "Ramona" is the result. This was in New York in the winter of 1883-84, and while thus engaged she wrote her publisher that she seemed to have the whole story at her fingers' ends, and nothing but physical impossibility prevented her from finishing it at a sitting. Alluding to it again on her death-bed, she wrote : " I did not write Ramona;' it was written through me. My life-blood went into it, — all I had thought, felt, and suffered for five years on the Indian question."

The report made by Mrs. Jackson and Mr. Kinney is grave, concise, and deeply interesting. It is added to the Appendix of this new edition of her book. In this California journey Mrs. Jackson found the materials for " Ramona," the Indian novel, which was the last important work of her life, and in which nearly all the incidents are taken from life. In the report of the Mission Indians will be found the story of the Temecula removal, and the tragedy of Alessandro's death, as they appear in "Ramona." — *Boston Daily Advertiser.*

Mrs. Jackson's Letter of Gratitude to the President.

The following letter from Mrs. Jackson to the President was written by her four days before her death, Aug. 12, 1885: —

To GROVER CLEVELAND, *President of the United States :*

Dear Sir, — From my death-bed I send you a message of heartfelt thanks for what you have already done for the Indians. I ask you to read my "Century of Dishonor." I am dying happier for the belief I have that it is your hand that is destined to strike the first steady blow toward lifting this burden of infamy from our country, and righting the wrongs of the Indian race.

With respect and gratitude,

HELEN JACKSON.

Sold by all booksellers. Mailed, post-paid, on receipt of price, by the publishers,

ROBERTS BROTHERS, BOSTON.

HELEN JACKSON'S WRITINGS.

BITS OF TRAVEL. Square 18mo. Cloth, red edges. Price, $1.25.

The volume has few of the characteristics of an ordinary book of travel. It is entertaining and readable, from cover to cover; and when the untravelled reader has finished it, he will find that he knows a great deal more about life in Europe — having seen it through intelligent and sympathetic eyes — than he ever got before from a dozen more pretentious volumes. — *Hartford Courant.*

BITS OF TRAVEL AT HOME. Square 18mo. Cloth, red edges. Price, $1.50.

The descriptions of American scenery in this volume indicate the imagination of a poet, the eye of an acute observer of Nature, the hand of an artist, and the heart of a woman.

H. H.'s choice of words is of itself a study of color. Her picturesque diction rivals the skill of the painter, and presents the woods and waters of the Great West with a splendor of illustration that can scarcely be surpassed by the brightest glow of the canvas. Her intuitions of character are no less keen than her perceptions of Nature. — *N. Y. Tribune.*

GLIMPSES OF THREE COASTS: California and Oregon; Scotland and England; Norway, Denmark, and Germany. 12mo. Cloth. Price, $1.50.

Helen Hunt Jackson has left another monumental memorial of her literary life in the volume entitled "Glimpses of Three Coasts," which is just published and includes some fourteen papers relating to life in California and Oregon, in Scotland and England, and on the North Shore of Europe in Germany, Denmark, and Norway. The sketches are marked by that peculiar charm that characterizes Mrs. Jackson's interpretations of Nature and life. She had the divining gift of the poet; she had the power of philosophic reflection; and these, with her keen observation and swift sympathies and ardent temperament, make her the ideal interpreter of a country's life and resources. — *Traveller, Boston*

BITS OF TALK ABOUT HOME MATTERS. Square 18mo. Cloth, red edges. Price, $1.00.

"Bits of Talk" is a book that ought to have a place of honor in every household; for it teaches, not only the true dignity of parentage, but of childhood. As we read it, we laugh and cry with the author, and acknowledge that, since the child is father of the man, in being the champion of childhood, she is the champion of the whole coming race. Great is the rod, but H. H. is not its prophet! — MRS. HARRIET PRESCOTT SPOFFORD, *in Newburyport Herald.*

Sold by all booksellers. Mailed, post-paid, on receipt of price, by the publishers,

ROBERTS BROTHERS, BOSTON.

HELEN JACKSON'S WRITINGS.

POEMS : Complete, comprising " Verses by H. H."
and " Sonnets and Lyrics." Square 18mo. Red edges,
price, $1.50; white cloth, gilt, $1.75.

Shortly after the publication of " Verses " Ralph Waldo Emerson
walked into the office of the publishers and inquired for the " Poems of
H. H." While he was looking at it the attendant ventured to remark
that H. H. was called our greatest woman poet. " The 'woman' might
well be omitted," was the only reply of the Concord philosopher. He
was then engaged in compiling his poetical anthology (Parnassus), in the
preface to which he says : " The poems of a lady who contents herself
with the initials H. H. in her book, published in Boston (1874), have a
rare merit of thought and expression, and will reward the reader for the
careful attention which they require."

JUVENILES.

BITS OF TALK, in Verse and Prose. For
Young Folks. Square 18mo. Cloth. Price, $1.00.

It is just such a book as children will enjoy, made up as it is of a variety of
attractive reading, short stories, fairy tales, parables, and poems, with here and
there a chapter of good advice, given in such a taking way without a bit of
goody talk, that the children will find it pleasant to take, little as they like advice
after the usual fashion. — *Worcester Spy.*

NELLY'S SILVER MINE. A Story of Colorado Life. With Illustrations. 16mo. Cloth. Price,
$1.50.

" Nelly's Silver Mine " is one of those stories which, while having the noble
simplicity and freshness whereby the young are captivated, is full of a thought
and wisdom which command for it the attention of all. — *Philadelphia Inquirer.*

CAT STORIES. Containing " Letters from a Cat,"
" Mammy Tittleback and her Family," and " The Hun-
ter Cats of Connorloa," bound in one volume. Small
4to. Cloth. Price, $2.00 ; or, each volume separately,
$1.25.

The subject is attractive, for there is nothing children take a more real in-
terest in than cats : and the writer has had the good sense to write neither above
nor below her subject. The type is large, so that those for whom the book is
intended may read it themselves. . . . For details we must refer all interested
to the story itself, which seems to us written with admirable verisimilitude. —
London Academy.

---

*Sold by all booksellers. Mailed, post-paid, on receipt of
the price, by the publishers,*

ROBERTS BROTHERS, Boston.

HELEN JACKSON'S WRITINGS.

RAMONA. A Story. 12mo. Cloth. Price, $1.50. (80th thousand.)

The Atlantic Monthly says of the author that she is "a Murillo in literature," and that the story "is one of the most artistic creations of American literature." Says a lady: "To me it is the most distinctive piece of work we have had in this country since 'Uncle Tom's Cabin,' and its exquisite finish of style is beyond that classic." "The book is truly an American novel," says the *Boston Advertiser*. "Ramona is one of the most charming creations of modern fiction," says CHARLES D. WARNER. "The romance of the story is irresistibly fascinating," says *The Independent*. "The best novel written by a woman since George Eliot died, as it seems to me, is Mrs. Jackson's 'Ramona,'" says T. W. HIGGINSON.

ZEPH. A Posthumous Story. 12mo. Cloth. Price, $1.25.

Those who think that all the outrage and wrong are on the side of the man, and all the suffering and endurance on the side of the woman, cannot do better than read this sad and moving sketch. It is written by a woman ; but never, I think, have I heard of more noble and self-sacrificing conduct than that of the much-tried husband in this story, or conduct more vile and degrading than that of the woman who went by the name of his wife. Such stories show how much both sexes have to forgive and forget. The author, who died before she could complete this little tale of Colorado life, never wrote anything more beautiful for its insight into human nature, and certainly never anything more instinct with true pathos. A writer of high and real gifts as a novelist was lost to the world by the untimely death of Mrs. Jackson. — *The Academy, London.*

BETWEEN WHILES. A Collection of Stories. 12mo. Cloth. Price, $1.25.

Mrs. Helen Jackson's publishers have collected six of her best short stories into this volume. Most of them appeared in magazines in the last year or two of her life. "The Inn of the Golden Pear," the longest and by far the strongest of them all, is, however, entirely new to the public.

Outside of her one great romance ("Ramona"), the author has never appealed to the human heart with more simple and beautiful certainty than in these delightful pictures. — *Bulletin, San Francisco.*

Mrs. Helen Jackson's "Little Bel's Supplement," the touching story of a young schoolmistress in Prince Edward's Island, is not likely to be forgotten by any one who has read it. The high and splendid purpose that directed the literary work of ' H. H.," and which is apparent in nearly everything that came from her pen, was supported by a peculiar power, unerring artistic taste, and a pathos all her own. This charming tale and one about the Adirondacks and a child's dream form part of the contents of this posthumous volume, to which, on her death-bed, she gave the beautiful title "Between Whiles." It is worthy to be placed alongside of her most finished pieces. — *Commercial Advertiser, New York.*

MERCY PHILBRICK'S CHOICE. 16mo. Cloth. Price, $1.00.

HETTY'S STRANGE HISTORY. 16mo. Cloth. Price, $1.00.

These two stories were originally published anonymously, having been written for the "No Name Series" of novels, in which they had a large popularity.

Sold by all booksellers. Mailed, post-paid, on receipt of price, by the publishers,

ROBERTS BROTHERS, BOSTON.

www.ingramcontent.com/pod-product-compliance
Lightning Source LLC
Chambersburg PA
CBHW021459090426
42739CB00009B/1790